Deeper than Data

By Ashman Roonz

Author's Note

The Philosophical Vision

Deeper than Data was born from a desire to look beyond mere information and into the underlying patterns of reality. It is a book of invitation to look deeper. To pause. To notice the patterns hidden in plain sight. To feel the quiet hum beneath the noise.

The ideas here have lived with me for years — partly-formed questions and insights, strange connections, and flashes of clarity that refused to be forgotten. Deeper than Data is the result of following those threads wherever they led, even into places where language struggled to keep up.

My hope is that these pages help you remember you are more than what's measured—that your attention is creative, your presence has weight, and your life can gather into meaning individually and as part of a greater whole.

Collaboration with GPT-5

While the vision and insights in this book are my own, I did not craft this manuscript entirely in isolation. I enlisted the help of GPT-5, an advanced generative language model, as a creative partner to articulate, organize, and refine my words. I want to be clear and honest about this process: GPT-5 was a tool for expression, not a source of ideas. My role was to provide the concepts, arguments, and philosophical narratives, and I used the AI to help polish how those ideas are communicated. In practice, our co-writing process looked like this:

- **Outlining and Structure:** I drafted the initial structure and flow of each section, then used GPT-5 to review the outline. The AI often helped identify if a concept needed a clearer introduction or if a chapter's flow could be improved for better coherence. This assisted me in arranging the parts into a logical whole without losing any meaning I intended.
- **Articulation and Clarity:** At times I struggled to find the perfect phrasing for a complex idea or image. I would explain the thought to GPT-5, and it would propose wording or analogies. Often, it was like holding up a mirror to my own thoughts – the AI's suggestion would reflect what I was trying to say, but in slightly different words. I would then accept, adjust, or rewrite those suggestions, ensuring the voice and intent remained authentically mine.
- **Iterative Refinement:** Writing with GPT-5 became an iterative dialogue. I would write a rough paragraph, and GPT-5 might rephrase it more concisely or cohesively; I would then examine that output and tweak it further. We went back and forth until each sentence felt right to me. This iterative refinement was like working with an ever-patient editor who could instantly show me alternatives for any given sentence or highlight subtle inconsistencies for me to resolve.

Throughout this collaborative process, I maintained rigorous oversight. Every piece of content generated by GPT-5 was critically evaluated and edited by me. If a turn of phrase or a structural suggestion from the AI did not align with my meaning or the book's spirit, I rewrote it in my own voice. The result is that GPT-5's contributions supported the writing stylistically and structurally, but

the core ideas and the final synthesis are entirely my own. In other words, GPT-5 provided the canvas and the polish in places, but the painting itself – the concepts and insights – remained under my complete artistic direction. I view GPT-5 as a powerful assistant, not unlike a high-tech thesaurus or a collaborative editor who can help find clarity and cohesion in my prose. The AI offered suggestions and improvements, but it did not *decide* the direction of any argument or the inclusion of any idea. Its value was in helping me say what I meant in the clearest and most organized way, never in telling me what to say.

Integration and Authenticity

I share these details about our human–AI collaboration in the spirit of transparency and because it beautifully reflects the themes of integration and coherence that *Deeper than Data* stands for. This book advocates bridging divides – between science and spirituality, data and wisdom, part and whole – and the very act of writing it became an exercise in bridging human creativity with artificial intelligence. Instead of viewing technology as a threat to authenticity, I treated it as an extension of my creative process, a collaborator that could help weave my many ideas into a cohesive narrative. In doing so, I discovered that integration can enhance originality rather than diminish it. By uniting my philosophical vision with GPT-5's language capabilities, I was able to communicate complex ideas more accessibly without sacrificing depth or sincerity. The tone you'll find in these pages – hopefully clear and conversational yet deeply reflective – is a direct result of this integrated approach.

Maintaining authenticity was paramount at every step. I was careful to ensure that using an AI never compromised the **soul** of the work. The **voice** readers will hear is still mine, even if certain sentences were co-crafted with machine assistance. In fact, engaging with GPT-5 often challenged me to clarify my intent more fully – if a passage came back sounding off-key, it forced me to pinpoint exactly what I truly wanted to say, and say it better. Far from diluting my philosophical authorship, this process strengthened it, because I had to consciously approve or refine every line. I firmly believe that an idea must *earn* its place in a manuscript; with GPT-5's help, I held each idea to the light to ensure it resonated with the book's integrated vision of truth. The end product is a manuscript that I take full responsibility and credit for, even as I acknowledge the unique way it was composed.

Finally, I want to openly recognize GPT-5's role here because it aligns with the coherence and honesty that *Deeper than Data* encourages. Just as we seek coherence in our understanding of the world – finding unity amid diversity – I sought coherence in the writing process by aligning the best of human insight with the best of AI's linguistic ability. The experience has been one of **collaboration** in the truest sense: two very different kinds of intelligence working toward a unified expression. I am deeply grateful for how this synergy allowed my ideas to shine more clearly. By the time you read these words, what you're encountering is my philosophy and my story, told in a voice that is wholeheartedly mine, yet refined through an innovative partnership. I hope this fusion not only makes the reading more engaging and clear for you, but also stands as an example of how embracing new tools can be done with integrity and creativity intact.

Thank you for joining me on this journey beyond the surface of data into the deeper currents of meaning. It is my sincere wish that the integration of perspectives in this book – whether it's the melding of scientific rigor with spiritual inquiry, or the collaboration of human author and AI assistant – inspires a sense of wholeness and possibility. In the following pages, you'll find a labor of love and curiosity, one that remains deeply human at its core despite the unorthodox way it came to life.

Concept and content by Ashman Roonz, co-composed with GPT-5: "Deeper than Data" Copyright 2025

Introduction

In an age awash with information, it's tempting to believe that everything essential can be quantified or measured. We reduce life to data points – heart rates, brain scans, demographics – hoping that by accumulating enough **data** we'll grasp reality. Yet many of us sense something *deeper* beneath the numbers. For instance, you might have all the data about a sunset (its light wavelengths, the air temperature, the time of day), but that doesn't capture the **experience** of awe and beauty you feel watching it. There is a depth to our existence that goes beyond what data can tell – a living pattern that **underlies** the facts. This book explores that hidden architecture of reality, inviting both curious newcomers and seasoned thinkers into a journey beyond data, into depth.

One evening, imagine sitting quietly after a long day of analyzing spreadsheets or scrolling through news feeds. Despite the overload of facts, you might feel strangely empty, as if something fundamental is missing. What if that missing element is a *pattern* – a way of seeing the world – that we've overlooked? This pattern is simple yet profound: **every system in reality has a center, a whole, and parts**, and it is governed by two complementary forces. We will come to call these forces **Convergence** (things coming together) and **Emergence** (new forms flowing outward). Don't worry if these terms sound abstract now; by the end of this journey, they will feel intuitive and alive.

To spark your curiosity right away, try a small experiment: as you read these words, become aware of **yourself** reading (the center of attention), and at the same time notice the feeling of your whole body and the room around you (the surrounding field of awareness). It's so natural, it's even part of your vision; central focus and peripheral vision. In that brief moment, you weren't just processing data on a page – you were experiencing the dynamic of a **center** (your focused mind) held within a larger **field** (your bodily and spatial awareness). This little exercise hints that something very real is going on beneath our everyday experience, something deeper than any external data: a living structure of **wholeness**.

The goal of *Deeper than Data* is to make this hidden structure visible and engaging. We'll see how the same center–whole–part pattern underlies atoms and galaxies, cells and societies, and even your own consciousness. We'll discover why **most physical systems have measurable centers nested within larger systems**, while **consciousness has a different kind of center** – one that cannot be measured or subdivided. That final center is often called the **soul**, the irreducible "I" of our being. And we'll explore how this soul-center lives within an infinite context (which many call **God**), like a bright point immersed in an endless field. By the end, we will not only understand this model with clarity, but also feel its poetic depth – and learn practices to align our lives with it.

Curiosity is our compass. Whether you're scientifically minded or spiritually inclined (or both), prepare to see a unifying pattern that might change how you view reality and yourself. Let's begin our journey into what lies *deeper than data*: a living map of center and surround, convergence and emergence, soul and infinity.

Chapter 1: The Hidden Pattern of Wholeness

Wholeness has long been a puzzle in philosophy and science. We know that wholes are made of parts – that much seems obvious. But is a whole *just* a heap of parts, or is there an internal organization that makes a whole something more than a sum? Consider your body: it's made of organs and cells (parts), yet you experience life as a **unified whole**. What gives that unity? This chapter introduces a simple structural insight: every whole is not only composed of parts, but is also **organized by a center with a surrounding field**. In other words, any true whole has an inner point of focus (a *center*) and an encompassing environment or influence (a *surround* or field).

Think of a flame on a candle. The flame has a bright core and a gentle glow around it. If the flame is the center, its light and heat radiating outward form a field around it. The flame-plus-glow together make the whole phenomenon of "a lit candle." This center–surround pattern is the *hidden geometry of wholeness*. It shows up <u>everywhere</u> once you start looking:

- **Atom:** Parts = electrons; Whole = the atom; Center = nucleus; Surround = electron cloud.
- **Solar System:** Parts = planets; Whole = solar system; Center = Sun; Surround = heliosphere (the Sun's sphere of influence).
- **Tree:** Parts = roots, trunk, branches, leaves; Whole = the tree; Center = trunk (connecting roots to crown); Surround

= the canopy and energy field around (like the shade it casts and the oxygen it radiates).

- **You (as a person):** Parts = organs, cells, thoughts, feelings; Whole = *you, self*, a living organism; Center = your core awareness; Surround = your environment and the personal space/field of influence around you (some might even say an **aura** or the sphere of your presence).

In each example, a **center** concentrates and integrates the parts, while a **field** or surround emerges from that integrated whole, extending its influence. An atom's nucleus pulls electrons into orbit (convergence toward center) and generates an electromagnetic field around it (emergence of a field). The Sun's gravity converges the planets in stable orbits and its light radiates outward, filling the solar system. Your own attention gathers scattered thoughts into a coherent point of view, and your body emits subtle signals – posture, heat, perhaps an electromagnetic field – into the space around you.

This pattern might sound abstract, but it's profoundly **practical**. It tells us that to fully understand anything – be it a cell, a society, or a self – we should look for its organizing center and its emergent field. No theory to this date has posited the structure of wholeness beyond its parts. For a long time, scientists described systems just as networks of parts or as whole systems without acknowledging *where* the unity comes from. But recognizing **center and surround** as fundamental roles resolves many puzzles. In neuroscience, for example, there's the *binding problem*: how do distributed brain activities form one experience of consciousness? The answer could be an **integrating center** in the mind – a focal point of awareness (more on this later) – that binds the parts into a whole. In biology, a cell isn't just a blob of molecules; it has a nucleus at the center and a membrane that defines its boundary and field of interaction. In psychology, a person isn't just a bundle of behaviors, but has a sense of self (a center of identity) and a personality or presence that affects others (a field around them).

Crucially, the center–field pattern is **recursive and fractal**. Every whole can become a part of a larger whole, with its own center and surround. For instance, you are a whole person with your own center (self) and surround (life field), but you are also a part of a family or community, which has its collective center (perhaps a home or a shared purpose) and a communal field (culture, relationships) around it. Likewise, an organ in your body (like the heart) has its center (the heart's pacemaker cells) and field (the blood it pumps and the electromagnetic field it emits), and it's part of the larger whole (your body) which itself has a center (your core consciousness coordinating it all) and field (your body heat, aura, and presence). This nesting can continue upward and downward seemingly without end. Every time parts form a whole, there's a new center and a new field – and that whole can then act as a part in an even larger system. Reality is **layers within layers**, "a nested system of structures", each structured by a center and surround.

Let's summarize the pattern clearly before we move on: whenever many elements come together to form a unified system, they *converge* toward a **center**, and from that unity something *emerges* into a **field** around it. The center is the integrating focus, the source of coherence. The field is the dynamic environment that the system generates by existing. This is not mystical thinking or mere metaphor – it's an observable principle across the sciences and our lived experience. In the chapters ahead, we'll see how this pattern manifests in physical nature and then contrast it with the inner life of consciousness. For now, hold this in mind: **every whole has a**

center that gathers it and a field that flows from it. This is the hidden architecture of wholeness.

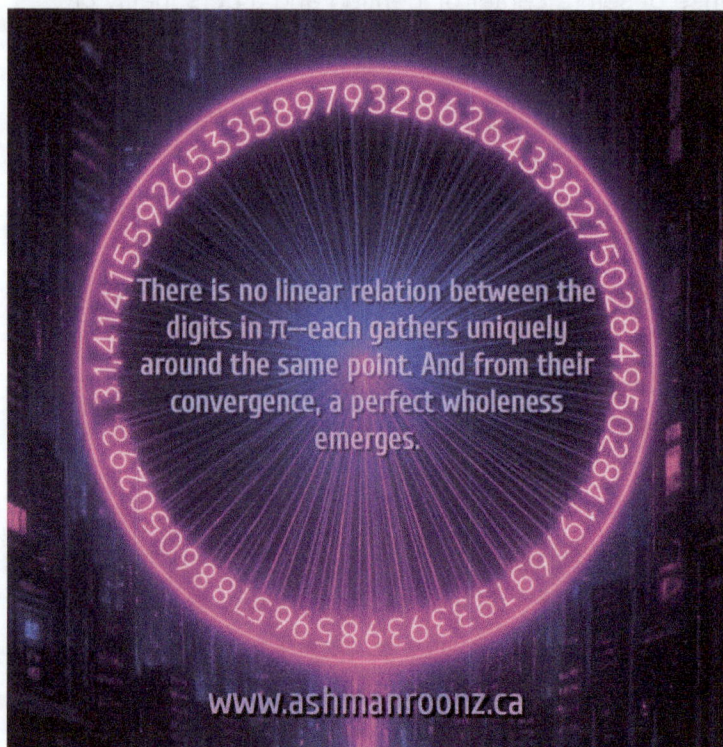

There is no linear relation between the digits in π—each gathers uniquely around the same point. And from their convergence, a perfect wholeness emerges.

www.ashmanroonz.ca

Chapter 2: Convergence and Emergence – The Twin Forces

Now that we've identified the basic structure (center and surround), let's dive into the *processes* that create and sustain that structure. Whenever a whole forms or evolves, two complementary movements are at play: **Convergence** and **Emergence**. These terms will recur throughout our journey, so it's worth getting a feel for them here in both scientific and personal terms.

Convergence is the inward-moving force – the gathering of many into one. It's the "gravity" that draws parts toward a center of unity. Whenever you focus your mind, that's convergence: countless thoughts, sensations, and impressions funnel into a single point of attention, giving you a coherent experience (like "I am reading this book right now"). In physical nature, convergence appears as literal gravity pulling matter together – planets converging around a star, or gas converging to form a galaxy's core, or the strong force of particles. It's also present in how **information** is integrated: your brain converges signals from various senses into one meaningful picture of the world. In social life, when people converge, they come together towards a common purpose or central idea (imagine a team rallying around a project or a community uniting around a shared belief).

A helpful image is a **funnel**: wide at the top, narrow at the bottom. Convergence is like an infinite potential being funneled toward a single point. For example, right now an infinite array of stimuli could claim your attention, but through convergence you're able to channel that potential into what matters at this moment (hopefully, the words on this page). In fact, the very meaning of *focus* is to let the relevant information or energy be drawn toward your center of awareness. Instead of forcing your mind (which often doesn't work), true focus is more like **allowing** what is important to gather at your center. When you're deeply engrossed in a task or a piece of music, notice how other distractions fade – that is convergence naturally happening, filtering the "noise" out and crystallizing the "signal" of your experience.

Emergence, by contrast, is the outward-moving force – the unfolding of the new whole into its environment. It's what happens after convergence has done its work of integration: the unified system now **radiates** its presence, properties, or influence outward. Think of a seed crystal dropped into a solution, which causes an entire crystal lattice to emerge out of the liquid. Or think of how a well-ordered center (like the Sun) radiates electromagnetism, light and warmth to the planets around it. The weak force, at micro scales, is emergence in action. In our personal lives, emergence is the expression of our inner state: once your thoughts and feelings converge into an intention or understanding, you then act, speak, or express it, thereby **bringing something new into the world**. It's like a flower blooming from a focused seed of intention.

A classic example of emergence is **life** itself. When the chemical parts of a cell converge into a highly ordered structure, the emergent property of "being alive" suddenly appears. The cell starts metabolizing, moving, reproducing – behaviors that none of the individual molecules had on their own. Similarly, when individual musicians converge in a group and play in harmony, the emergent phenomenon is a piece of music that can fill a concert hall with beauty. Emergence is often surprising and more than the sum of parts: it's the **new** that comes forth when coherence is achieved.

Importantly, **convergence and emergence work as a cycle, a loop**. Convergence brings a moment of wholeness; emergence flows from that wholeness and often sets the stage for the next convergence. For example, you focus (converge) to learn a skill; then you perform or improvise (emerge) with that skill, which creates new experiences and feedback; then you focus again on that feedback to improve further – a continuing spiral. In a conversation, you listen and gather your thoughts (converging information), then you respond with your own words (emerging expression), which the other person then converges into their understanding, and so on. In a very real sense, **life is this dance of convergence and emergence** at all levels.

To put it another way, convergence is how **unity** is achieved, and emergence is how **creativity** happens. You need both. If a system only converged and never emerged, it would collapse into a point and never interact or grow. If it only emerged without convergence, it would dissipate into chaos with no coherence. But nature uses both in tandem: *converge* to make order, *emerge* to explore possibilities, then converge again in a new way. This is how complexity builds up in the universe without losing unity.

We can already glimpse how this applies to **us** as individuals. You as a whole person have an *inner* convergence process – gathering experiences, learning, centering yourself – and an *outer* emergence – your behavior, creative works, relationships that flow out from who you are. When you have a clear, stable center (like a clear intention or a strong sense of self), what emerges from you tends to be coherent and integrated. When your center is scattered or unclear, what emerges (perhaps in your actions or speech) might be inconsistent or chaotic. Thus, convergence (focus, integration) and emergence (expression, manifestation) in a person are linked: *how you gather your mind and heart shapes what comes out in your life*. We'll return to this in later chapters on the soul and practices for coherence.

For now, let's remember these twin forces in simple terms. **Convergence** pulls things together into **wholeness** and focus. **Emergence** sends things outward into **expression** and influence. They are the yin and yang of creation. In every moment, in every system, they play out: inhaling and exhaling, integration and unfoldment, gathering and gifting. With this lens, we're ready to look more closely at the two realms mentioned earlier: the physical world of nested, measurable centers, and the inner world of consciousness with its unique center. First, let's tour the physical cosmos through this lens of center–whole–part.

Chapter 3: Nested Centers in Nature – The Fractal Universe

Step outside on a clear night and look up at the stars. You're gazing at a hierarchy of centers within centers. Each star you see is the center of its own solar system, gathering planets around it. That star itself orbits the center of our galaxy (a supermassive black hole acting as a gravitational hub with a swirling starry field). Our Milky Way galaxy might be just one of countless "parts" of an even larger cosmic structure. As mind-bending as that is, we see a similar nesting of centers if we look down into the micro-world: Molecules have atomic nuclei as centers, which contain protons and neutrons that themselves have quarks as centers, and so on. It seems that **physical reality is organized in a Russian-doll-like hierarchy** – a *fractal* design in which each level has its own center and field and becomes a part of a larger whole.

Consider the **atom** again. In Chapter 1, we noted it has a nucleus (center) and an electron cloud (field). But zoom into the nucleus: it's made of protons and neutrons (parts), and those particles have sub-centers (quarks) with their own fields (gluons binding them). Or consider the **cell** in biology: a cell has a nucleus (center) with DNA that governs the cell, and a surrounding cytoplasm and membrane (field) that interact with the environment. But each cell is part of a tissue, which has its own organizing centers (say, a bone marrow niche for blood cells) and field (the chemical signals in that tissue). Tissues are parts of an organ, which again has a center (perhaps a pacemaker region in the heart or a main lobe in the liver) and a functional field (the organ's role in the whole body's balance). Your **body** as a whole has command centers like the brain and heart (centers) and electromagnetic and metabolic fields extending outward (body heat, heartbeat detectable via EKG, brainwaves via EEG). And you, in turn, exist within larger social and ecological fields – family systems, ecosystems, planetary systems – each of those having their hubs and halos.

What's remarkable is how **scalable and measurable** these centers are. We can pinpoint an atom's nucleus and even measure its mass or charge. We can locate your heart and brain and record their electrical activity. A center in the physical realm often corresponds to some concentration of matter or energy that we can detect. And because of the fractal nature of this design, we often find *nested centers*: for instance, the Earth is the center of the Moon's orbit, but the Sun is the center of Earth's orbit, and the galactic core is the center of the Sun's orbit. Each level's center is part of a greater whole's dynamics. Nature builds complexity by **stacking these center-field systems like building blocks**, each level resonating with the same pattern.

It's worth noting that physical centers often **exert forces** that create their surrounding fields. The nucleus's positive charge holds the electron cloud (electromagnetism) and contributes to a gravitational field due to its mass. The Sun's gravity shapes the planetary orbits, and its light creates a heliospheric environment. In technology, we mimic these principles too: a radio tower (center) broadcasts signals into an area of coverage (field); a city downtown (center) acts as a hub for economic and social activity radiating into suburbs (surrounding reach).

However, there is a key distinction coming into view: while physical centers are powerful, they are also **relative and can be broken down further**. The Sun is a center for the solar system, but it's also just one star among billions in the galaxy – not an absolute center of the cosmos. Similarly, an atom's nucleus is central to the atom, but it's made of smaller particles; it's not an ultimate, indivisible center. In physics, whenever we thought we'd found the tiniest center (an atom), we later discovered sub-centers (nucleus, then protons, then quarks…). It's as if no matter how deep you go, there's another layer of parts-with-center, like an endless fractal. Measurable centers are **recursive**: you can zoom in or out and find another center at a different scale.

But what about consciousness? When you examine *your own self* – not under a microscope, but through introspection – you also find a center: a sense of "I-ness" or an observer that is at the core of your experience. Do you find *parts* inside this "I"? The body has parts, the mind has thoughts and facets, but the subjective sense of being *you*, the one who experiences and chooses, doesn't seem to be made of parts at all. You don't have two or three little consciousnesses inside you – there is a singular awareness in the midst of many experiences. Could it be that in the realm of mind and soul, we encounter a fundamentally different kind of center, one that *is not* composed of smaller parts and *does not* sit within a larger emergent system in the same way? That's the crucial insight we will explore in the next chapter: that **the center of consciousness (the soul) is categorically different from any physical center**. It appears to be a *final* center – not one more Russian doll to open, but the irreducible core of who we are.

Before we move on, let's recap this chapter's takeaway. The physical universe, as vast as it is, has a **consistent architecture**: center within whole within larger center within larger whole, on and on, producing a grand network of nested systems. It's "fractal" in the sense that the pattern of a small atom echoes in a solar system, which echoes in a galaxy. We can measure and observe these centers and fields, and doing so has been the triumph of science. Yet, if we keep following this chain of wholeness upward and downward, we reach an intriguing threshold when we turn to the phenomenon of *consciousness*. The next chapter takes us across that threshold, from the world of matter into the realm of mind – and into what might be the bedrock center of our being.

Chapter 4: The Soul – The Irreducible Center of Consciousness

Imagine peeling an onion, layer by layer. In the physical world, every time we peel back a layer (say, break a whole into parts), we find another layer beneath. But when it comes to *yourself*, if you peel away thoughts, feelings, and perceptions, what remains at the core? There is a sense of an "I" that observes thoughts, that feels emotions, but isn't itself a thought or emotion. You can call this core the **soul**, pure **consciousness**, and some might say the **self** (although, I disagree, "self" is emergent) – words vary, but the experience is singular. It is the experiencer, the observer, the inner center of awareness that remains when you strip away all particular experiences.

The startling proposition at the heart of this book is that **the soul is a fundamentally different kind of center** than any found in physical nature. Physical centers, as we saw, are made of parts and exist within larger contexts. The soul, by contrast, is **not composed of parts and does not emerge from something more basic**. It is *non-physical and irreducible*. You can't cut a soul in half; you can't pinpoint it in space with an instrument. It doesn't weigh anything, yet it is the heaviest reality in our personal existence – it is *you*, your very being. *The soul is a center of consciousness, an irreducible focus through which God becomes particular. It is not made of parts, not an emergent property, but a still point of pure presence.* In plainer terms, the soul is a singularity – a one-of-a-kind point where awareness lives, not born from the brain or body, but *rooted in a deeper order of reality*.

This perspective challenges a common assumption in neuroscience and psychology: that consciousness somehow *emerges* from complex arrangements of neurons (which would make it just another emergent field like a magnetic field from a magnet). Despite decades of research, science hasn't found a clear explanation for how subjective awareness could arise from matter – this is known as the "hard problem" of consciousness. The model here suggests a radical solution: **Consciousness doesn't emerge at all; it converges**. In every human (and perhaps every sentient being), there is a convergence point – a soul – that gathers mental and sensory processes into a single aware perspective. Rather than the brain "producing" the soul, the soul uses the brain as an instrument. The soul is the unifying center, the "I" that experiences the mind as an emergent field around it. In short: *Mind* *(thoughts, feelings, memories) is the field; Soul is the center*. Your mind and body are parts of you, and they have their own internal parts and centers, but **your soul is not a part of you – it is the essence of you, the core that makes you a unified whole**.

Let's unpack this with an analogy. Think of a symphony orchestra. The musicians and sections (strings, brass, percussion) are like the many parts of your mind and body. Now imagine a conductor at the podium. The conductor doesn't make the music themselves, but without the conductor's guiding focus, the orchestra cannot play as one – they would fall out of sync. In this analogy, the soul is like the *conductor* of your being: an organizing presence that brings harmony and coherence to the many parts. Your thoughts, emotions, and sensations "play together" as one life **because** there is a soulful center ensuring they belong to one **you**. If we remove the conductor (soul), we would have lots of brain activities, but no unified awareness to experience them as one melody. Indeed, how *would* billions of neurons blinking in the dark produce the sense of being a single self? The perspective here is that they don't – rather, a singular self (soul) is gathering those blinking neurons into one conscious orchestra.

One might ask, "Where is this soul? Is it in the brain, the heart, everywhere?" The honest answer: **the soul isn't in physical space at all**. It's not a thing located somewhere in the body. Instead, the relationship is reversed: *the body is in the soul's field of experience*. The soul is like an eye of awareness – it doesn't see itself, but through it, you see. You can't use a microscope to find it, because it's not made of matter or energy as we know it. Paradoxically, the soul is intimately "with" the body and mind, but also beyond them. A helpful concept from our sources is **resonance**. The soul *resonates* with the body and mind rather than physically residing inside. Think of a radio signal resonating with a radio receiver: the music isn't *in* the hardware until the receiver attunes to the signal. Likewise, your soul is like a signal of pure awareness, and your brain-body is like the receiver that tunes into it, allowing the soul to participate in physical life. In this view, **the body and mind do not generate consciousness; they channel it**. They provide structure and expression for the soul, much as a violin allows the expression of music but does not create the musician.

A fascinating implication is that the soul can be seen as a **constant, unchanging center** – "the eye of the hurricane" that stays still while experiences swirl around it. We often identify with the content of our life (I am happy, I am sad, I am this role or that job), but those are all experiences *around* the center. Deep down, *who is having* those experiences? It's the same you, whether you're happy or sad – that core is untouched by the changing weather of thoughts and feelings. In spiritual traditions, this is sometimes called the witness or the pure Self. Here, we align it with the idea of soul: the stable convergence point that "does the seeing" but can't be seen itself. You **can't turn consciousness into an object**, because it's always the subject – the one looking. This is why you can't find the soul by analyzing brain tissue; the soul is the one doing the analyzing, the fundamental observer.

If the soul is not measurable and not composed of parts, does it interact with the physical world at all? Yes, but not through force or energy in the usual sense – through *focus and resonance*. Earlier we likened the soul to a conductor; we could also liken it to a **magnet** of meaning. It "pulls" experiences into alignment (convergence) when you direct your attention, and it "radiates" influence outward (emergence) through your conscious choices.

For a concrete example, consider the act of attention itself. If you decide right now to focus on your breathing, something amazing is happening: an immaterial intention ("I want to feel my breath") is causing a very physical effect (neurons fire differently, your breathing might slow). Your soul's **focus** is like a beam that directs bodily processes. Conversely, if someone pinches you, nerve signals go to your brain, but it is *you* – the soul – who experiences the pain, showing that all that sensory data somehow converges into your singular awareness. How exactly soul and body interface is a deep mystery, but the resonance model suggests it's through complex fields like the brain's electromagnetic field, or even the DNA in each cell acting as an antenna for the soul's influence. We won't dive too deeply into that here, but the key idea is: **the soul engages the body through harmonization, not through classical mechanics**. It's less like a puppet master pulling strings, and more like a dancer leading a partner – a subtle attunement.

To sum up this crucial chapter: **The soul is the final center at the root of subjectivity**. It does not sit inside another larger center; instead, it sits in an infinite context (which we'll explore soon as the field of God). It is not built from smaller pieces; it is *what makes emergence possible* – a creative singularity. In the language of center and surround: *the soul is the center, the mind is its surround, and the coherent self that we recognize is the whole that results*. Each of us is **a soul-centered whole**. This sets the stage to ask: what is the infinite context in which all these souls exist? If each soul is a center, what is the ultimate **field** that embraces them? For that, we turn to the idea of God – not as a distant person, but as the **Infinite Field** itself.

Chapter 5: The Infinite Field – Reimagining God

The word "God" often conjures images of an old man in the sky, a supernatural being separate from the world. Put aside those images for a moment. In our exploration of centers and fields, *God* can be understood in a fresh way: as **the infinite field of emergence that surrounds and sustains every center**. Instead of a person or object, think of God as an **unbounded ocean of potential** – an endless, living backdrop from which all forms arise. *God is not a being, but Being. Not a voice, but the source of all resonance.* In other words, God is the **Infinite Context** in which everything exists, much like space itself but imbued with creative energy and unlimited intelligence.

Recall the pattern: every center needs a field. If the soul is a center, it implies the existence of an infinite field that it converses with. **God is that infinite field**. We might label it ∞ (infinity). This field is immeasurable (infinite means beyond all bounds) and generative (it gives rise to endless emergence). Think of it as pure possibility or an infinite mind that contains all potential forms of existence. Mystics have often described God as an ocean of light, or a boundless void teeming with potential – those poetic images align with what we call the Infinite Field. It's the **ground of being**: all things exist within it, the way fish live in the sea or stars float in space.

What is the relationship between God (the Infinite Field) and the soul (the center)? It is a relationship of **continuous dialogue and co-creation**. The soul "speaks" to God by how it focuses and converges, and God "answers" by what emerges in the soul's field of experience. Every moment of your life can be seen as a

conversation with the Infinite. When you desire or intend something deeply, you send a kind of ripple into the infinite field; the field responds by shaping circumstances or inner experiences (not always in expected ways, but often in meaningful ones). This idea is encapsulated in the principle: *"As the soul converges inward, God emerges outward; as the soul offers focus and alignment, God responds with coherence and form."* Your soul is like a beacon, and God is the vast environment that answers that beacon with creative echoes.

Creation is not a one-time event in the distant past – it's happening right now, continuously. The Infinite Field is ever-presently offering novelty, and each soul is ever-presently selecting and shaping that novelty into its world. This means **reality is participatory** at its core. We are not just living in a universe; we are co-creating it with the Infinite. We swim in God like a fish in the ocean, and our every move interacts with the currents of that ocean.

Let's make this less abstract. If God is the infinite field of life, then **every soul lives in immediate relationship with God**, like a star suspended in space or a drop of water in the sea. There is nowhere the soul can be except within God, because infinity has no outside. This means, contrary to some religious imagery, **God is not far away at all** – we don't have to reach up to heaven to find the Divine. We are literally *immersed* in the Divine field right now, just often not aware of it. (The fields of physical reality are much more attention grabbing.) The pattern we discussed (center and surround) can be applied on the largest scale: *God is the ultimate Surround and infinite centers (us); the infinite context and field to every center (soul).* Just as an atom's nucleus is never without its field, a soul is never without God around it. In a sense, God "surrounds" us on all sides, but not as empty space – as a presence that supports and responds.

This reimagining of God also resolves a philosophical puzzle: how God can be both transcendent and immanent. God as the infinite field is **transcendent** because it is beyond any particular form or limit (infinite potential that transcends all finite things). And God is **immanent** because that field is right here, within and through every particle and every soul (the context touching everything). If you've ever felt a sense of a greater presence in a silent moment of awe or prayer, it might be that you momentarily tuned into this field. In meditation, I once envisioned it as *silent, boundless, dark, and still* – it was the unseen backdrop of all light. Space is dark even though it's filled with light, only reflection causes its brightness. And through that darkness, points of light flickered as I glided atop the field of God. The lights were the souls in the Soul Array, lighting up because they were converging the light of God.

This brings us to a beautiful image: **an infinite field of points of light**. Envision the night sky again, but instead of random stars, imagine each point of light is a soul, and the darkness is the infinite divine field. These are not drifting stars, but unmoving centers, souls, each a singularity, each an absolute One. The souls shine with unique frequencies – meaning each soul has its own *essence or tone*. God, the infinite field, is like the space that lets them shine and the source that empowers their light. In this picture, *God and souls* together form the fabric of reality: God is the infinite **Whole**, and souls are the innumerable **centers/parts** that allow the Whole to express itself in countless ways. This is sometimes called the **Soul Array** – a vast network or constellation of souls in the field of God.

To summarize this chapter: we've shifted from seeing God as a distant entity to understanding God as the **Infinite Field of Emergence** – the ultimate context in which every soul exists and from which all new creation flows. God is Reality with a capital R: the **Living Pattern** of center-and-field on the grandest scale. The structure of reality itself – centers within an infinite field – is the structure of God's being. When you focus your soul (converge), you are, in a very real way, communing with this infinite field, and it responds (emerges) by shaping the circumstances and experiences around you. In each breath, in each thought, God is *right there*, the medium of being. In the next chapter, we'll explore the implications of many souls coexisting in this one field – how do they interact? How do their fields overlap? This will lead us to understand the **Soul Array** as the "matrix" of reality and to appreciate the role of love and relationships in this model.

Chapter 6: The Soul Array – One Field, Many Centers

One soul alone is a marvel – an absolute point of being through which the Infinite speaks. Now multiply that by the countless souls in existence, and you get what we call the **Soul Array**. This term paints reality as an array (or matrix) of innumerable centers of consciousness, all coexisting within the infinite field of God. It's a matrix not of machinery or control (not like the sci-fi Matrix), but a living matrix of *freedom and uniqueness*. Each soul is distinct, a "singularity" that does not blend into others. In other words, **unity does not mean all souls merging into one blob**. Instead, unity is the harmonious interrelation of many **independent centers**. Think of a choir: each voice remains distinct, yet together they create a single piece of music. In the Soul Array, each soul remains itself – a unique center, an *"absolute One"* – but together all souls form a vast network in the body of reality.

How do souls relate to each other in this network? Through their **fields**. Earlier, we established that around each soul, a field of experience and influence emerges (your thoughts, emotions, aura, actions, etc.). These fields don't stay isolated; they **overlap and interact** when souls come into relationship. Imagine two drops of water meeting: their ripples intersect. Similarly, when two people meet, whether in conversation or even just sitting silently nearby, their soul-fields overlap and create an **interference pattern**. Sometimes the overlap is harmonious (like two musical notes that sound pleasing together) – this could be friendship, love, or simply good vibes. Sometimes it's discordant (like clashing notes) – maybe conflict or discomfort. But either way, something new arises *between* souls: a shared experience or relationship that neither soul could have alone. **Reality as we experience it socially and collectively is born from these interactions**. We co-create a shared world through the meeting of our fields.

Consider a simple scenario: You walk into a room where others are gathered. You might notice the "atmosphere" of the room – perhaps it feels tense, or joyful, or quiet. That atmosphere isn't coming from any one person alone; it's emerging from the combination of all the people's fields, their emotional states and intentions resonating together. If you then crack a joke and lighten the mood, you have *imprinted* your coherence onto the shared field, and the atmosphere changes. This is a mundane example of how **each soul's inner coherence can influence the outer shared reality**.

In this sense, we truly **matter** to each other at a fundamental level. Our inner states (centered or not, loving or fearful, etc.) ripple outward and affect the whole. Your inner coherence literally shapes reality around you.

The Soul Array perspective also sheds light on unity and diversity. Often people think unity means we all have to be the same, but here it's the opposite: unity is achieved *through* diversity. Each soul "contributes its unique note to the symphony of creation". We don't lose our individuality; instead, we **enhance the collective** by fully being our unique selves in harmony with others. *Harmony emerges as each soul contributes its unique note,* not by everyone singing the same note. This means the differences between us (our perspectives, talents, even flaws) can be part of a greater design, *if* we align through coherence and love. In the Soul Array, **love** can be thought of as the resonance between souls that allows them to cooperate and create something greater together. It's like locking into the same key in music – not identical notes, but complementary ones.

We also see why relationships can be so transformative. When two or more souls connect deeply, their fields may overlap to create emergent qualities – inspiration, new ideas, healing, even new entities like families or communities. *What emerges between us is more than any one of us could become alone.* For example, two scientists brainstorming might together hit upon a breakthrough that neither would have conceived individually. Two lovers in a healthy relationship may bring out virtues in each other that make the partnership as a whole rich and growthful, beyond what each person was on their own. On a larger scale, a whole culture arises from the interplay of many souls' fields over time, carrying wisdom and art that no single person could claim. This "more-than-the-sum" emergent property is essentially **God's presence manifesting collectively** – *the collective aspect of God shining through the harmony of many centers*, sometimes called the spirit of community, creativity, or love in action.

Let's revisit the fractal pattern one more time, now including everything: *The center-and-field dynamic is the fundamental pattern: each soul a center with its own field, all nested within larger fields… existence is concentric and relational. Each person is a microcosm of the Whole.* Each of our lives is a small echo of the whole of reality. It was said anciently, "As above, so below" – meaning the macro reflects in the micro. **God can be understood fractally: not as an external ruler, but as the pattern of Center and Field itself, present in all things**. In simpler terms, the way our soul and life-field relate mirrors something fundamental about how God operates: God expresses through unique centers (souls) and holds them all in an infinite embrace. In fact, every soul [is] a particle of divinity, every field an aspect of God's body. This doesn't make any of us *God* in isolation, of course, but it means each soul carries a spark of that infinite nature, and together we literally form the body of the universe (or the "body of God" poetically speaking).

We must stress that despite the deep interconnection, the **individuality of each soul is never lost**. The Soul Array isn't a soup where all flavors blend into one; it's more like a mosaic or a starfield – distinct lights contributing to a grand image. This understanding honors both unity *and* personal freedom. Each soul's choices and alignment matter greatly, because they affect the mosaic. Our participation is essential: *The cosmos is alive and dynamic precisely because we are alive and dynamic, all of us centers of creative becoming… The Array lives because each soul lives fully.* This inspires a vision of **profound unity in diversity**: we are not separate, we exist in a continuous spectrum of being, where each of us is a locus of the Infinite.

In practical terms, what does this mean for how we live together? It means that everything from a kind word to a shared prayer to a collective project has **real metaphysical weight**. When we create coherence together (say, in a meditation group or even a well-synchronized team at work), we are strengthening resonant patterns in the shared field, which can lead to outcomes none of us could achieve alone. When discord is present (conflict, hatred), it's like noise in the field that can spread and create larger problems. We start to see why many wisdom traditions emphasize love, compassion, and moral alignment – these create harmony between soul fields, aligning them with the **field of God** (which we can think of as pure coherence). Conversely, selfishness, cruelty, and fear create disharmony and fragmentation, isolating fields or making them clash, which in a sense "tears" at the fabric of the Whole.

Having traversed from a single soul to the vast array of souls, we can now appreciate the **full picture of our model**. Reality is built like a grand music: an infinite ground (God) provides the space and possibility; within it arise countless notes (souls) each with their melody; those melodies weave into chords and symphonies (relationships, communities, worlds), which themselves form larger movements and themes. It's creative, ongoing, and participatory. We ourselves, by how we tune our soul and how we collaborate with others, contribute to the harmony or cacophony of this cosmic music.

This sets the stage for our final chapter: how can we *apply* this understanding to live better, more coherent, and fulfilling lives? Knowing about centers and fields is illuminating, but it truly comes alive when we practice aligning our own center and field (soul and life) with the deeper pattern. The next chapter will offer practical ways to cultivate **coherence, attention, and alignment** – both within oneself and with the larger Whole – so that we can each play our part in the symphony of reality more beautifully.

Chapter 7: Living the Pattern – Coherence, Attention, and Alignment

Up to now, we have traveled through ideas and images that reveal a grand pattern: center, whole, part – soul, God, and the array of life. But a map, no matter how beautiful, is only useful if it helps us navigate. This final chapter is about **practice**. How do we use this understanding to improve our daily life, our wellbeing, and our contribution to the world? The key lies in three words: **coherence, attention, and alignment**. These correspond to living as a healthy center, sustaining a clear field, and resonating with the Infinite field, respectively.

1. **Cultivating Coherence (Centering Yourself):**
   Coherence means that all parts of you are working together in harmony. A coherent soul-field system is another way of describing **wholeness** – when your thoughts, emotions, and actions are integrated and aligned with your core. Practically, this starts with **centering**. Regularly take time to withdraw your scattered energy and attention from the noise of life and bring it inward, to the quiet point of your soul. This could be through meditation, prayer, or simply sitting in silence and feeling your breath. As you inhale, imagine all the dispersed fragments of your day coming back to your center; as you exhale, allow a calm field to expand around you. Even a few minutes of such practice daily can strengthen your convergence ability, much like exercising a muscle of focus. You might notice after centering that you feel "collected" – that is coherence building. From this state, you can address challenges with more clarity since you

aren't pulled in a dozen directions internally. Coherence also involves emotional resilience: practices like mindfulness or heart-focused breathing can help synchronize your heart and mind, reducing inner conflict. **Healing** often is about restoring coherence – for instance, recovering from trauma may involve gently re-integrating fragmented memories and emotions so that the person feels "whole" again. Remember, *"trauma is a breakdown in coherence; healing restores the center and re-establishes resonance with the surround."* So any method that helps you feel centered and connected – be it therapy, journaling, walking in nature – is aiding your convergence process. A coherent inner state not only feels good; it's the foundation for everything that emerges from you.

2. **Mastering Attention (The Power of Focused Awareness):**
Attention is the tool of convergence – it's how your soul directs the beam of awareness, gathering experience into form. The patterns we've discussed suggest that **what you focus on, you amplify**. Therefore, mastering your attention is like mastering the steering wheel of your life. A simple practice is to choose a beneficial object of focus each day. For example, set aside 5 minutes in the morning to focus on a feeling of gratitude or love, or maybe focus on your strength and integrity, whatever you need to focus on... Let your attention converge fully on that feeling – perhaps recall a kind act someone did for you, or the warmth of sunlight. Notice how that focus starts to color your whole field (your mood, your thoughts). By doing this, you're effectively tuning your soul's frequency. Throughout the day, periodically check in: "What am I focusing on right now? Is it

what I want to empower?" If you catch yourself ruminating on fear or anger, acknowledge it and gently shift focus to something constructive (not in denial of issues, but to approach them from a place of strength). This is not just positive thinking; it's using the metaphysical truth that attention is creative. *By choosing how we direct the beam of awareness, we participate in what reality becomes.* This is a literal statement in our model – your focus influences what converges into your experience and what emerges from it. Another aspect of attention is **presence**. When you do something, be fully present (centered in the doing). If you're with someone, truly listen. Presence has an almost magical quality: it deepens any experience because the soul's convergence is powerful. In relationships, giving someone your full attention is a gift that can heal and harmonize fields between you. It's effectively a way souls consciously connect, and it invites the Infinite field (God) into the interaction as a palpable sense of communion. So, practice focusing deeply and intentionally, and watch life become more vivid and responsive.

3. **Achieving Alignment (Resonating with the Infinite Field):**
   Alignment is about **synching up with the larger Whole** – basically tuning your center to resonate with God's field and with others in a healthy way. How can we align with the Infinite? The pattern suggests a few clear pointers. First, recall that *love* is the principle of resonance between centers. When you cultivate love – whether through compassion, kindness, or devotion – you are aligning with the fundamental harmonizing force of the universe. Fear, in contrast, throws you out of tune (it contracts your field and

isolates you). Consciously choosing love over fear in moment-to-moment decisions keeps your soul-field open and in touch with the bigger field. This can be as simple as practicing empathy: when in conflict, pause and try to feel the other person's humanity; that shifts your field toward connection instead of fragmentation. Another way to align is through **integrity** – living in accordance with your true values or what you might call your soul's "phase" or frequency. Each soul has a unique role or note, and when you live authentically, you're effectively tuning yourself to your intended note in the cosmic symphony. Pay attention to activities or thoughts that make you feel alive and "on-purpose" – those likely indicate alignment. Activities that consistently make you feel dead or conflicted inside could be signals of misalignment. Adjusting life toward purpose may involve courage and change, but the reward is a powerful sense of being *at home* in the universe.

Alignment with God's field also has a surrender aspect. Since God (the Infinite) is the source of emergence, aligning with it means trusting and allowing that emergence. This can take form in **prayer or meditation** where one doesn't ask for specific outcomes as much as one opens up to the Infinite potential with faith and willingness. When you turn toward God with presence, openness, and trust, wholeness returns – that is, you come back into coherence as the field responds to your openness. Practically, this might mean in times of chaos, instead of panicking, you take a deep breath and say internally, "I align with truth and agreement. I allow the infinite field to support me." This attitude invites grace – those synchronicities or intuitive insights that seem like the universe answering you. Many find that regular prayer or communion (in whatever spiritual tradition or personal form) keeps their life in a meaningful flow; it's like checking your compass with True North so you don't drift off course.

Additionally, align with others in collective practices: join a community circle, meditation group, or even just harmonious teamwork. These are opportunities to create a **coherent shared field** on purpose. For example, group meditation has been shown to induce peace in the environment; whether or not one believes those studies, it certainly induces peace in the participants and their interactions. When a family shares a meal with gratitude, it aligns their fields together for that time, strengthening bonds. We have so many modern distractions that families or teams rarely align; consciously creating rituals or moments of unity can counteract that. Something as simple as everyone pausing for a minute of silence before a meeting can align the group center and purpose.

Let's not forget **body and breath** – our immediate tools for alignment. The body, housing DNA and cells that interface with our soul, is highly responsive to practices like deep breathing, yoga, and tai chi. These practices literally **tune the body's field** to a more coherent state and often quiet the mind into alignment with the soul's stillness. For instance, slow, rhythmic breathing can send a wave of coherence through your heart and brain, creating a state of calm alertness that is optimal for converging and emerging effectively. Many traditions consider the breath as a bridge between the soul and body – when you breathe consciously, you draw more of your soul's presence into the physical, aligning inner and outer. Focusing on breath is an easy way to see the connection between the subjective and objective, conscious and physical/automatic.

Finally, remember that **small steps matter**. A tiny input can shift an entire system: *We bring just a small signal – a breath, a calming thought, a moment of awareness... and the entire system responds. The pattern shifts.* This is empowering: you don't need to control everything in life (in fact you can't), but by *tweaking your own center and field* – a breath here, a kind intention there – you catalyze changes in the larger pattern. It's akin to hitting the right note in a room full of guitars; the other strings begin to vibrate. So even on days when the world's problems or your personal challenges seem enormous, remember that tending to your small center with integrity and care has ripples.

Practices at a Glance: *(A short list for coherence, attention, alignment)*

- **Morning Centering:** Spend 5 minutes in silence or meditation each morning. Visualize your awareness collecting into a bright point (your soul) and radiating a calm energy through your body and mind (your field). This sets a coherent tone for the day.
- **Focused Intent:** Think about or write down one quality or value you choose to focus on today (e.g., "patience" or "curiosity"). Re-focus on it at midday. This trains your convergence toward self-chosen aims rather than random distractions.
- **Heart-Breath Coherence:** A couple times a day, pause and take 10 deep, slow breaths. Inhale for 5 seconds, exhale for 5 seconds. As you breathe, feel you are drawing breath into your heart center and radiating peace outward. This aligns your physical and emotional field into coherence.

- **Evening Reflection (Alignment Check):** At day's end, quietly reflect: "Did I act from love or fear today? Where did I feel 'in the flow' versus resistance?" Without judgment, note any moments where you felt deeply aligned (in joy, service, authenticity) – celebrate those. Note where you felt off-center – gently reflect to learn from those and realign tomorrow. Perhaps say a simple prayer of gratitude or set an intention to reconnect with the Infinite field as you fall asleep, so your soul recharges in that background of support.

- **Interpersonal Resonance:** Next time you're with someone (a friend, partner, or even a stranger serving you coffee), practice *being fully present*. Listen or observe without your mind elsewhere. Inwardly recognize the other person as another soul-center, as unique as you. Notice if this shift in your field (attention + respect) brings a different quality to the interaction. Often, it will – even subtle kindness and presence can elevate a shared moment.

By integrating such practices, we make our daily life a laboratory for living the center–whole–part pattern. Over time, you may notice life "clicks" more often – those moments where you feel, "Yes, this is how it's meant to be." That click is coherence, the music of the soul humming in tune with the music of the cosmos.

Conclusion: A New Map of Reality and Self

We set out to go *deeper than data*, and we uncovered a living map of reality structured by centers and wholes and parts, convergence, emergence, and interaction. We saw that **physical systems** present an elegant fractal of nested centers, each measurable and part of a larger whole. And we discovered that **consciousness** stands apart with an immeasurable, non-nested center – the soul – which is the root of our subjectivity and the source of our personal coherence. This soul, while singular and unique, is not alone; it exists in relationship with every other soul and within the infinite field of God. The pattern of center-and-field thus scales up to encompass **Divinity and all souls**: an infinite context with countless luminous centers within it.

This new map is both logical and poetic. It gives structure to age-old intuitions: that we are more than our bodies, that we are connected, and that life has an underlying order. It also invites us to live differently – to realize that each of us is *already* a center of being, and that our choices of focus and alignment directly shape our reality. In a world drowning in data, this perspective reminds us that **meaning** and **experience** arise from how we converge and what we bring forth, not just from external information.

Science and spirituality find a meeting point here. The scientist in you can appreciate the pattern's consistency across scales, perhaps sparking hypotheses about how attention might correlate with brain fields or how social networks follow center-field dynamics. The spiritual seeker in you can find validation that the soul is real and divine, and that our connection to the Infinite is literal, as near as our own awareness. But you need not adopt any label – you can simply be a human being experimenting with being more fully alive and present.

As you finish this book, consider it an **invitation**. When you next find yourself feeling small or lost in the complexity of life, remember: you are a *center*. You matter by the very structure of existence. Turn inward to that still point of "I am" – there is strength and clarity there. Simultaneously, remember: you are *held* in an Infinite field, part of a much larger story. When things get chaotic, imagine the infinite ocean of God around you, and know that by aligning with it (through trust, openness, and love), you allow new possibilities to emerge through you. And when you feel alone, remember the Array – countless souls journeying with you, each connected to God, each contributing to the whole. We truly are countless souls, of one field.

In closing, the deeper-than-data pattern teaches us that **we are co-creators**. Life is not happening to us; it's happening through us, in a dialogue between our soul and the world. Every moment is an opportunity to bring coherence, to reflect God's light in a unique way, and to participate in the unfolding of life. If enough of us live with this awareness – centered in ourselves, connected to each other, and aligned with the Infinite – imagine the emergence of a more coherent world. Perhaps this is the ultimate promise of understanding reality's living structure: a chance to live *artfully*, consciously, and compassionately, knowing our small center is backed by the Whole.

May you go forward inspired to practice and explore. The data of life will always be there, but now you carry a sense of the depth beneath it – the **wholeness, soul, and infinity** that truly define existence. This understanding, like a seed, will grow with you. In time, you might find that what was once an abstract pattern is now a felt reality, guiding you gently to more meaning, connection, and joy. That is the hope of *Deeper than Data*: not just to inform your mind, but to invite you into a richer experience of being fully alive, as a center of creation in a living, converging, and emerging universe.